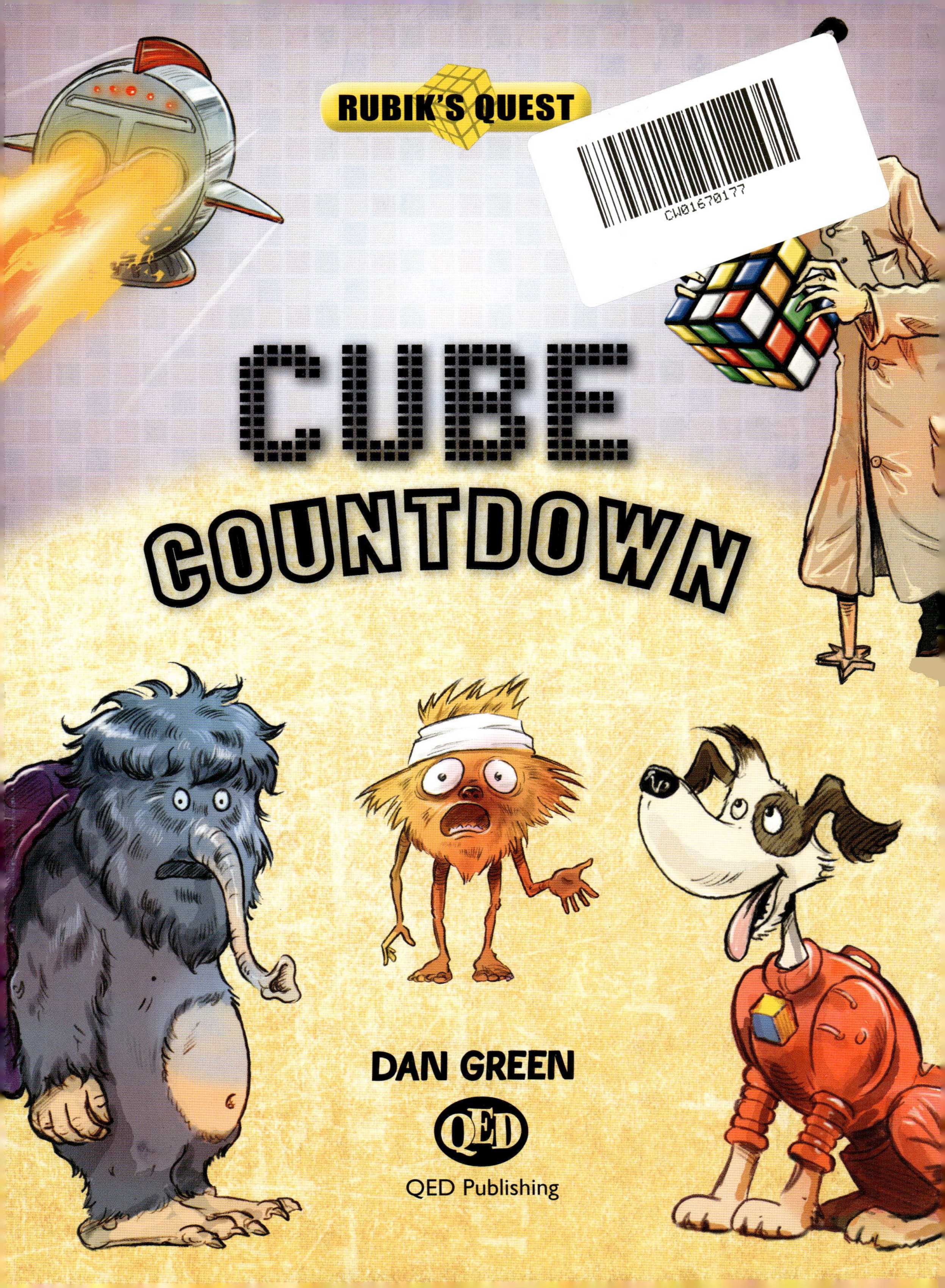
RUBIK'S QUEST
CUBE
COUNTDOWN
DAN GREEN
QED
QED Publishing

Cover Design: Rosie Levine
Illustrator: David Shephard
Editor: Amanda Askew
Designer: Punch Bowl Design
QED Project Editor: Ruth Symons
Editorial Director: Victoria Garrard
Art Director: Laura Roberts-Jensen

Picture credits: (t=top, b=bottom, l=left, r=right, c=centre, fc=front cover)

Bigstock Milissenta 9c 21c, 23r, 38t, 40t; homestudio 13t, 19c, 21b, 31br, 32c, 42t; Im Perfect Lazybones 21c, 28t, 31bl; Vasilius 30cb; krishnacreations 42cr; mike301 38b, 45t; **Shutterstock** Ruslan Semichev 26c.

First published in the UK in 2014 by
QED Publishing, A Quarto Group company
The Old Brewery, 6 Blundell Street, London, N7 9BH

www.qed-publishing.co.uk

A catalogue record for this book is available from the British Library.

ISBN 978 1 78171 863 6

Printed in China

How to begin your adventure

Are you ready for an awesome adventure in which you must solve mind-bending puzzles? Then you've come to the right place!

Cube Countdown isn't an ordinary book – you don't read the pages in order, 1, 2, 3... Instead you jump forwards and backwards through the book as you face a series of challenges. Sometimes you may lose your way, but the story will always guide you back to where you need to be.

The story begins on page 4. Straight away, there are questions to answer and problems to overcome. The questions will look something like this:

IF YOU THINK THE CORRECT ANSWER IS A, GO TO PAGE 37 ★

IF YOU THINK THE CORRECT ANSWER IS B, GO TO PAGE 11 ⟳

Your task is to solve each problem. If you think the correct answer is A, turn to page 37 and look for the matching symbol in red. That's where you will find the next part of the story. If you make the wrong choice, the text will explain where you went wrong and let you have another go.

The problems in this book are all about shapes, patterns, Rubik's Cube games and number puzzles – prepare to have your brain tested! To solve them, you must use your knowledge and common sense. Use a calculator to help you with complex maths problems. To help you, there's a glossary of useful words at the back of the book, starting on page 44.

ARE YOU READY?

Turn the page and let your adventure begin!

CUBE COUNTDOWN

You wake with a start and answer the phone.

Life is never dull as a Lieutenant at the Cube Command Centre, even on your day off! You and your dog Superflip quickly get ready and hurry to HQ.

GET READY TO **START YOUR ADVENTURE** ON PAGE 8

Afraid not! You need to look for a pattern in the numbers.

TURN BACK TO PAGE 19 AND HAVE ANOTHER TRY

Incorrect. The * symbol is not on the opposite side.

TRY AGAIN ON PAGE 36

33 moves would be a good score, but not the best.

HAVE **ANOTHER GO** ON PAGE 41

Mr Algebrains, Superflip and you cautiously leave the patrolship. Everywhere is quiet.

To enter the Cube Command Centre, you must input a randomly generated security code.

SECURITY CODE

How many different ways can a Rubik's Cube be arranged?

100,456,877
JUMP TO PAGE 19

43,252,003,274,489,856,000
GO TO PAGE 32

300,253
TURN TO PAGE 12

Phew! You're through the portal safely. The symbol you chose was the same shape as your ship, but upside down.

You fly through the Time Torus and land on a flat plain. There's a trail of star-shaped footprints outside, so you're definitely in the right place.

Hang on! Superflip looks odd... He's become two-dimensional and so have you! You tell Superflip to follow the footprint trail, while you take a look around.

Suddenly, a Stickman is pointing a spear at you.

Stickman ties your hands and puts you on a cross-shaped hoverboard. Quick-thinking Superflip hides behind a tree...

YOU JET TOWARDS A PALACE ON PAGE 39

Wrong! With one cut, you can only make a rectangle or an isosceles triangle (with two equal sides).

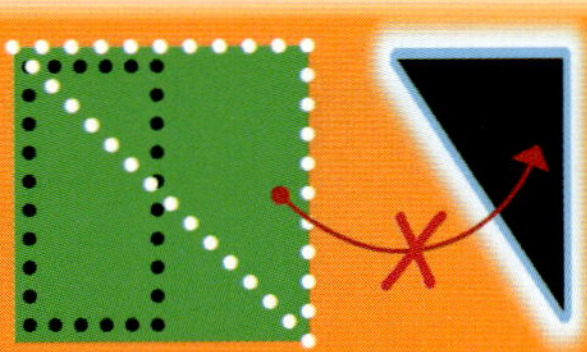

TRY AGAIN on PAGE 27

No, $\frac{1}{3}$ would be 0.33 kg.

THINK AGAIN on PAGE 33

When you arrive, the room is chaotic. People are running around and shouting. Commanders Korners and Vertex approach you. Vertex starts to speak.

What do you answer?

6 **TURN TO PAGE 19**

8 **FLIP TO PAGE 31**

12 **GO TO PAGE 42**

Correct! 2, 3, 5 and 7 can only be divided by 1 or themselves, so they are called prime numbers. You travel 17 km north to the Navigator's house.

You find him outside, pacing up and down. You ask him the way to Flatland.

You offer to help him in return for his navigation skills.

Which two fences should you reposition to turn five small squares into four?

FENCES A AND B
TURN TO PAGE 21

FENCES D AND F
GO TO PAGE 12

FENCES C AND E
FLIP TO PAGE 38

Wrong! You need that piece to make the square.

TRY AGAIN on PAGE 17

Wrong! If you cut the cheese into 11 pieces, you won't get $\frac{1}{4}$ for yourself.

TRY AGAIN on PAGE 14

9 Well done! You spotted the pattern.

$$7 \times 7 = 49$$

Think carefully before you answer.
You want to impress Korners.

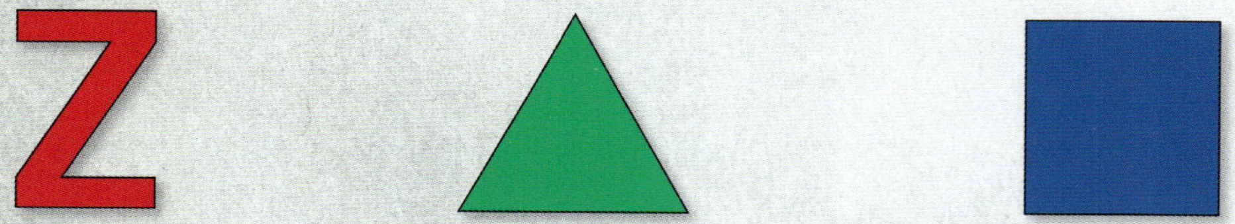

TURN TO PAGE 18 Z

HEAD TO PAGE 26 ▲

FLIP TO PAGE 37 ■

= That's not right! The = symbol is not on the opposite side.

TRY AGAIN ON PAGE 36

📎 Try again! Each snark eats $3 \times \frac{3}{4} = 2\frac{1}{4}$ kg every day.

GO BACK TO PAGE 20 AND HAVE ANOTHER TRY

48 Think carefully. It's true that each of the 6 faces has 8 stickers that move, which equals 48 stickers. But you're looking for how many cubies can move, not stickers.

TURN BACK TO PAGE 21 AND TRY AGAIN

Correct! After 48 rings, the door swings open and you step inside.

You enter a grand room. In the middle is the King of Flatland, holding what looks like a cube. Cubes can't exist in Flatland, so it must be the stolen Core!

You snatch it from the King, but soon realize it's just a mosaic tile. Frustrated, you throw it to the floor.

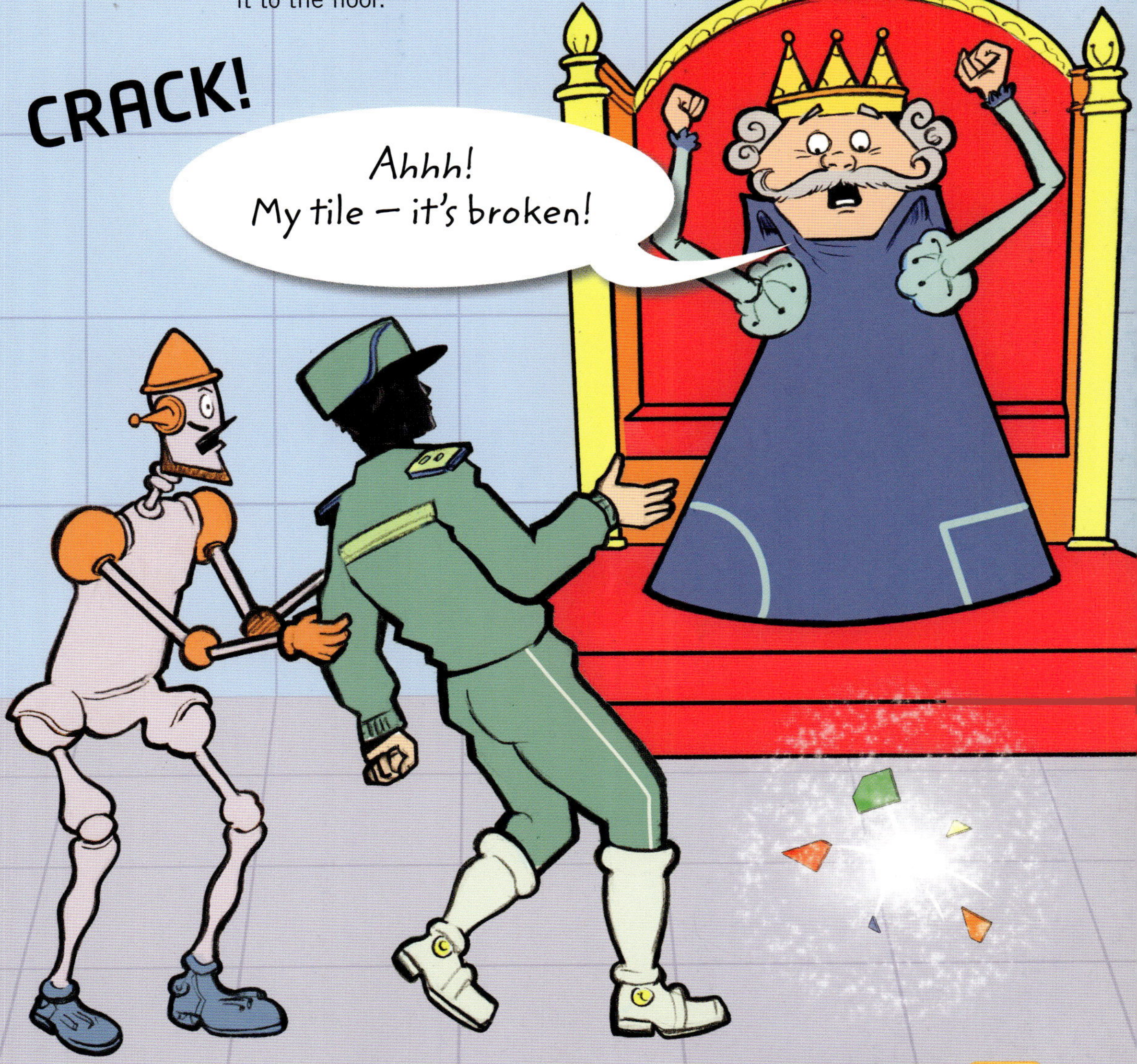

Before you know it, Stickman is dragging you out of the room.

SEE WHERE HE TAKES YOU on PAGE 27

Correct! Moving fences D and F leaves you with four equal-sized squares!

Happy again, the Navigator explains that the only way you can get to Flatland is by using the Time Torus, a portal on the edge of Planet Star.

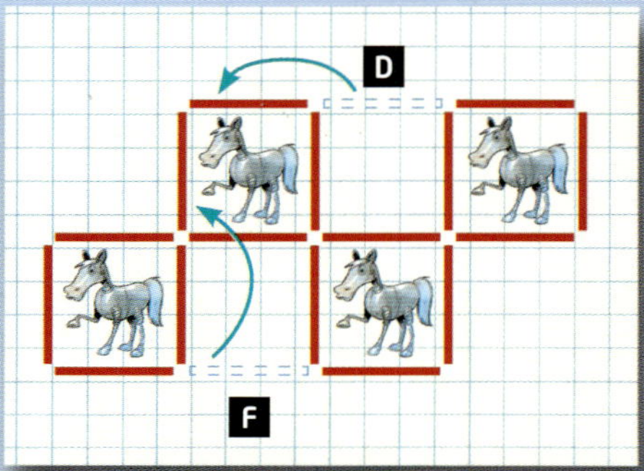

Which numbers do you type into your computer?

3, 3, 2, 5
GO TO PAGE 40

9, 2, 5, 3
HEAD OVER TO PAGE 28

7, 3, 3, 1
TURN TO PAGE 20

There are more arrangements than that.

GO BACK TO PAGE 5 AND **TRY AGAIN**

Afraid not! A nonillion is a pretty big number – 1 with 30 zeros after it. But it isn't the biggest number!

GO BACK TO PAGE 42 AND **TRY AGAIN**

$\frac{1}{2}$ is is wrong! Perhaps it's better to think of it like this:

$$\frac{1 \times 1}{4 \times 4} = ?$$

TURN BACK TO PAGE 31 AND **TRY AGAIN**

No, 4, 6 and 8 are not prime numbers because they can each be divided by 2, 3 or 4. A prime number can only be divided by 1 or itself.

TRY AGAIN
ON PAGE 30

Steady on! Although there are 6 faces, each with 9 stickers, equalling 54 stickers, there aren't 54 cubies.

TURN BACK TO PAGE 21
AND **TRY AGAIN**

You've got it!
5 + 8 = 13,
8 + 13 = 21
and
13 + 21 = 34.
Phew, that was a close call!

What do you answer?

NAUTILUS MOLLUSC SHELLS
TURN TO PAGE 38

SUNFLOWER SEED HEADS
GO TO PAGE 26

THE COCHLEA IN THE INNER EAR
FLIP TO PAGE 42

ALL THREE
HEAD OVER TO PAGE 21

A

Correct! You've spotted that **5 x 3 x 2 = 30**. You fly straight to Fraction Land. Phew – you're three-dimensional again!

You exit the ship and are nearly bitten by a hungry snark! As far as the eye can see, snarks are *eating the planet!* If you don't stop them, your ship will be next!

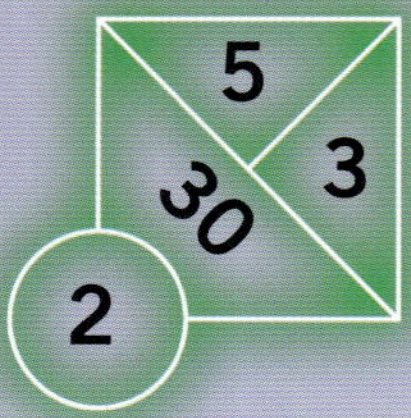

You decide to tempt the snarks with a slice of mooncheese each. If you keep $\frac{1}{4}$ of the cheese and there are nine snarks to catch, how much cheese will they each get?

$\frac{1}{9}$ GO TO PAGE 37

$\frac{1}{12}$ JUMP TO PAGE 33

$\frac{1}{11}$ TURN TO PAGE 9

Correct! $\frac{1}{4}$ of $\frac{1}{4}$ is $\frac{1}{16}$.
It's not the biggest slice of cake you've ever had, but it's better than nothing.

$$\frac{1 \times 1}{4 \times 4} = \frac{1}{16}$$

You explain to the snark-keeper that you're on a mission to find the missing Rubik's Core, and you were led here by an All-seeing Cube. You only have 14 hours left to find it, so you need to get a move on.

The snark-keeper tells you where to find the Patrol HQ.

YOU AND SUPERFLIP MAKE YOUR WAY THERE ON PAGE 42

It's impossible to move the pellets correctly in only 6 moves.

TURN BACK TO THE CONTROLS ON PAGE 24 AND **TRY AGAIN**

No, that's not right.

444 x 4 = 1776

TRY AGAIN ON PAGE 29

No, you've made a strange shape that won't fit through the square!

GO BACK TO PAGE 39 AND **TRY AGAIN**

Fantastic! If you cut the rectangle along its diagonal you will make a scalene triangle (with no equal sides). It fits the lock perfectly!

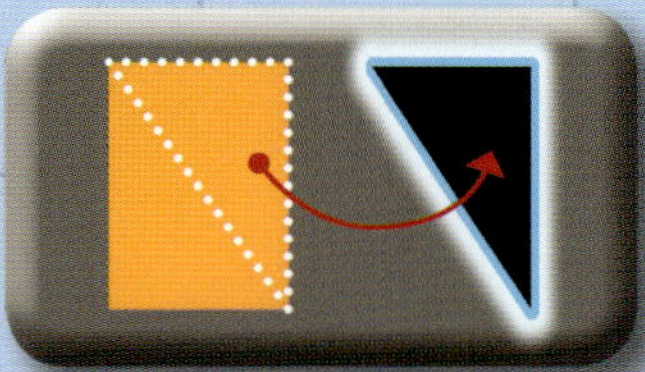

You need to find Superflip and follow the star trail. As you make your way towards the triangle door, you hear a yelp. Superflip?

You peek round the door of the grand hall. Superflip is in a cage next to the broken tile.

"LET HIM GO!" you shout.

You start putting the square tile back together, but there seems to be one piece too many. Which is the extra piece?

THE RED PIECE. GO TO PAGE 9

THE BLUE PIECE. TURN TO PAGE 43

THE PURPLE PIECE. FLIP TO PAGE 36

No! Although a centillion is enormous – 1 with another 303 zeros after it – it isn't the largest number.

GO BACK TO PAGE 42 AND HAVE ANOTHER GO

Well done! A Z-shape has a rotational symmetry of order 2. In a complete turn, there are two positions in which it looks the same.

That's not long! You'd better get a move on. Superflip races ahead of you and boards your spaceship.

YOU ROAR TOWARDS PLANET STAR ON PAGE 24

Illogiko's face darkens. Superflip yelps. You'd better fix the Cube before your canine pal meets a sticky end.

Which 2D shape will form a 3D Rubik's Cube?

SHAPE 1
TURN TO PAGE 28

SHAPE 2
HEAD TO PAGE 26

SHAPE 3
FLIP TO PAGE 41

Unbelievable, but the number is much bigger than that!

HAVE ANOTHER GO ON PAGE 5

Right! The cubies in the centre of each side are not shared between faces.

Good start. But how's your maths? Second question: Can you work out what number is missing from this pattern?

Look at how the numbers work together before you answer.

7
GO TO PAGE 27

4
TURN TO PAGE 5

9
HEAD OVER TO PAGE 10

Oh dear! That's not even close! Look closely – there are triangles of many different sizes. The trapdoor starts to creak...

QUICKLY, TRY AGAIN ON PAGE 43

Although they are prime numbers, 7 is not a factor of 90. Factors of a number are numbers that divide it exactly (e.g. 3 is a factor of 9).

TURN BACK TO PAGE 12 AND THINK AGAIN

That's right! 0.75 is the same as $\frac{3}{4}$. The snark-keeper has cheered up.

$9\frac{1}{2}$ KG
GO TO PAGE 10

$20\frac{1}{4}$ KG
JUMP TO PAGE 31

$27\frac{3}{4}$ KG
TURN TO PAGE 43

Correct! The shell, sunflower head and cochlea all feature the spiral.

As your destination is still some way off, you recline your seat and relax for five minutes. You ask Superflip and Mr Algebrains if they fancy doing some brainteasers.

Yes, I'll start! Look at this number pattern. Which of the following numbers also fit this pattern?

$$19 = 1 \times 9 + 1 + 9$$

38 TURN TO PAGE 28

11 TURN TO PAGE 41

29 TURN TO PAGE 23

No, moving fences **A** and **B** will give you 3 squares and an L shape.

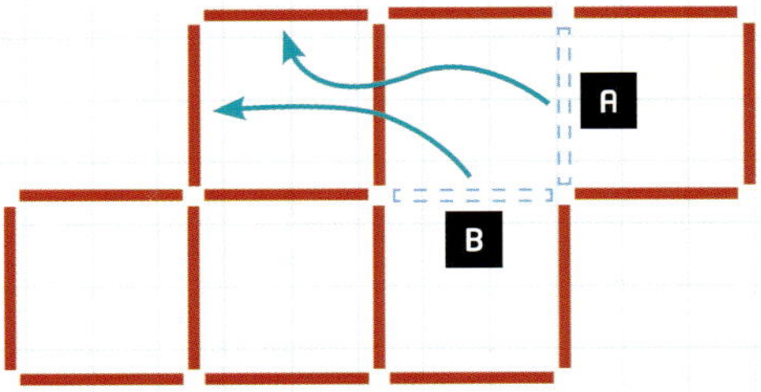

TRY AGAIN
ON PAGE 9

Exactly! So you've got to solve the Cube in 20 moves or fewer.

Certain cubies never change position. But how many cubies can move?

20 GO TO PAGE 34

48 TURN TO PAGE 10

54 FLIP TO PAGE 13

Well done! The **X** shape is on the opposite face. You can now ask the All-seeing Cube a question.

"Where is the stolen Rubik Computer's Core?"

The All-seeing Cube floats into the air, spins and an image beams out. A map!

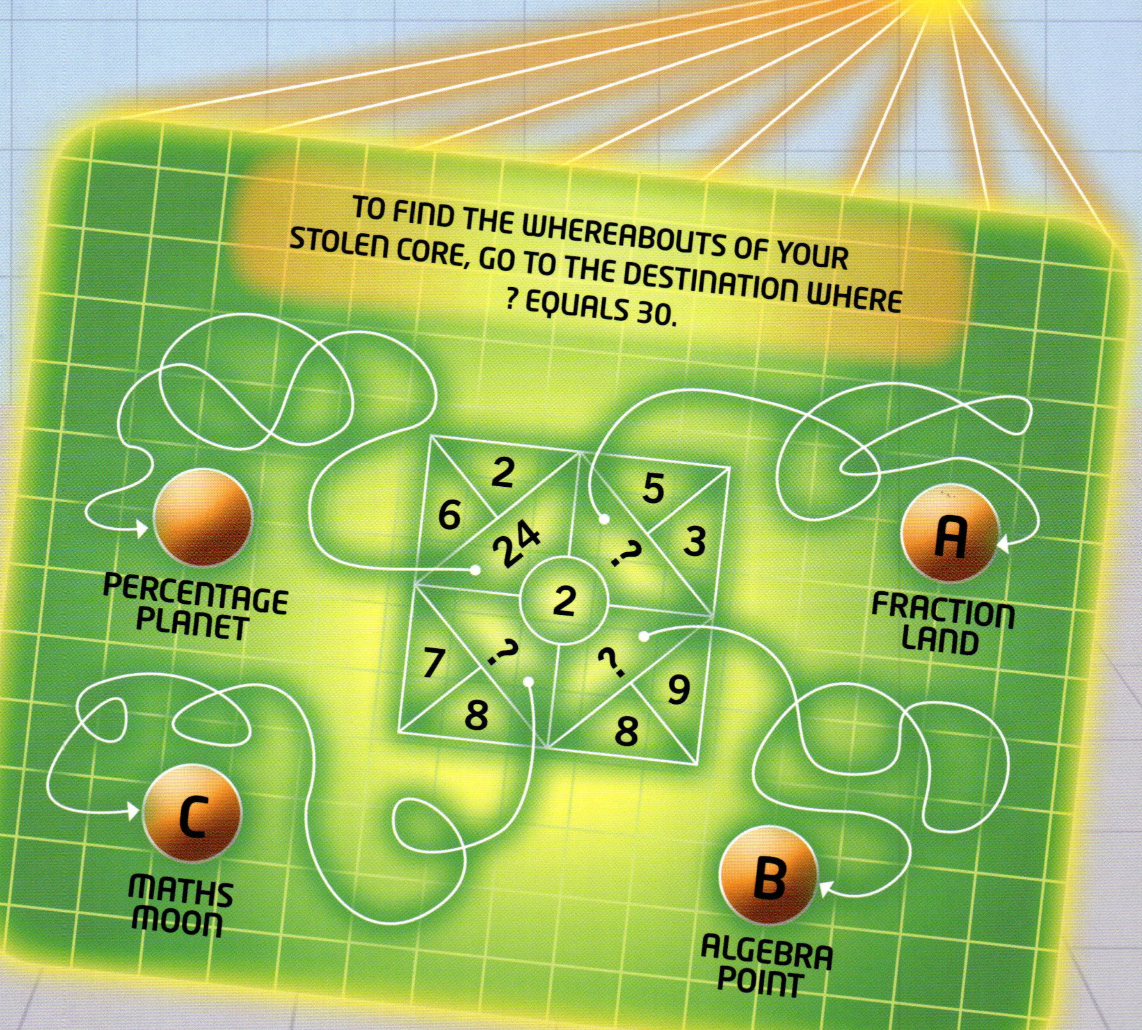

Look carefully at the number puzzle.
Which destination do you choose?

A
GO TO PAGE 14 **A**

B
FLIP TO PAGE 33 **B**

C
HEAD OVER TO PAGE 38 **C**

Wrong answer! Your ship isn't the right shape to fit through portal 3.

CHOOSE ANOTHER PORTAL ON PAGE 40

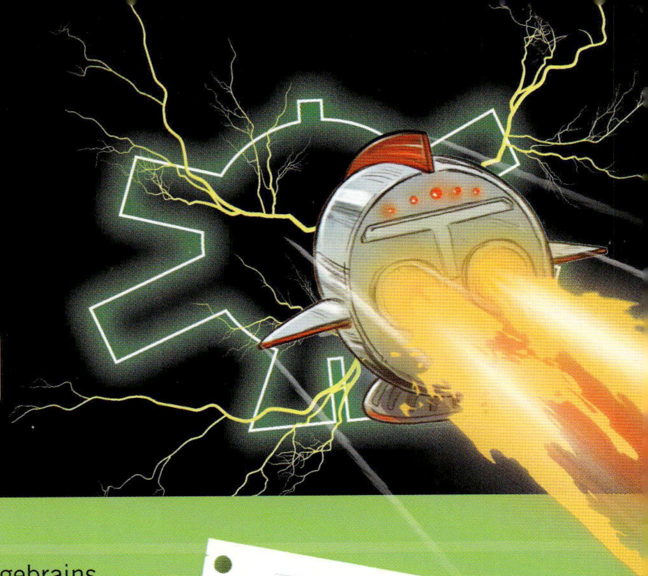

With a smug grin, you tell Mr Algebrains that the answer is 29. He looks crushed that you solved it so quickly.

$2 \times 9 + 2 + 9 = 29$

Before you challenge him to a brainteaser, you check in with the Cube Command Centre to let them know your progress. Vertex appears on the screen, looking frightened.

Never put off by danger, you reprogram the ship's coordinates.

YOU HEAD FOR THE COMMAND CENTRE ON PAGE 5

As soon as you land on Planet Star, you can see that something is wrong. An ear-piercing alarm is ringing and flashing lights nearly blind you!
Lieutenant, have you come to save us? The spaceport's power source is overheating. We have minutes to shut it down before we are all vaporized!
You weren't expecting that, but you have to help if you want to stay alive!

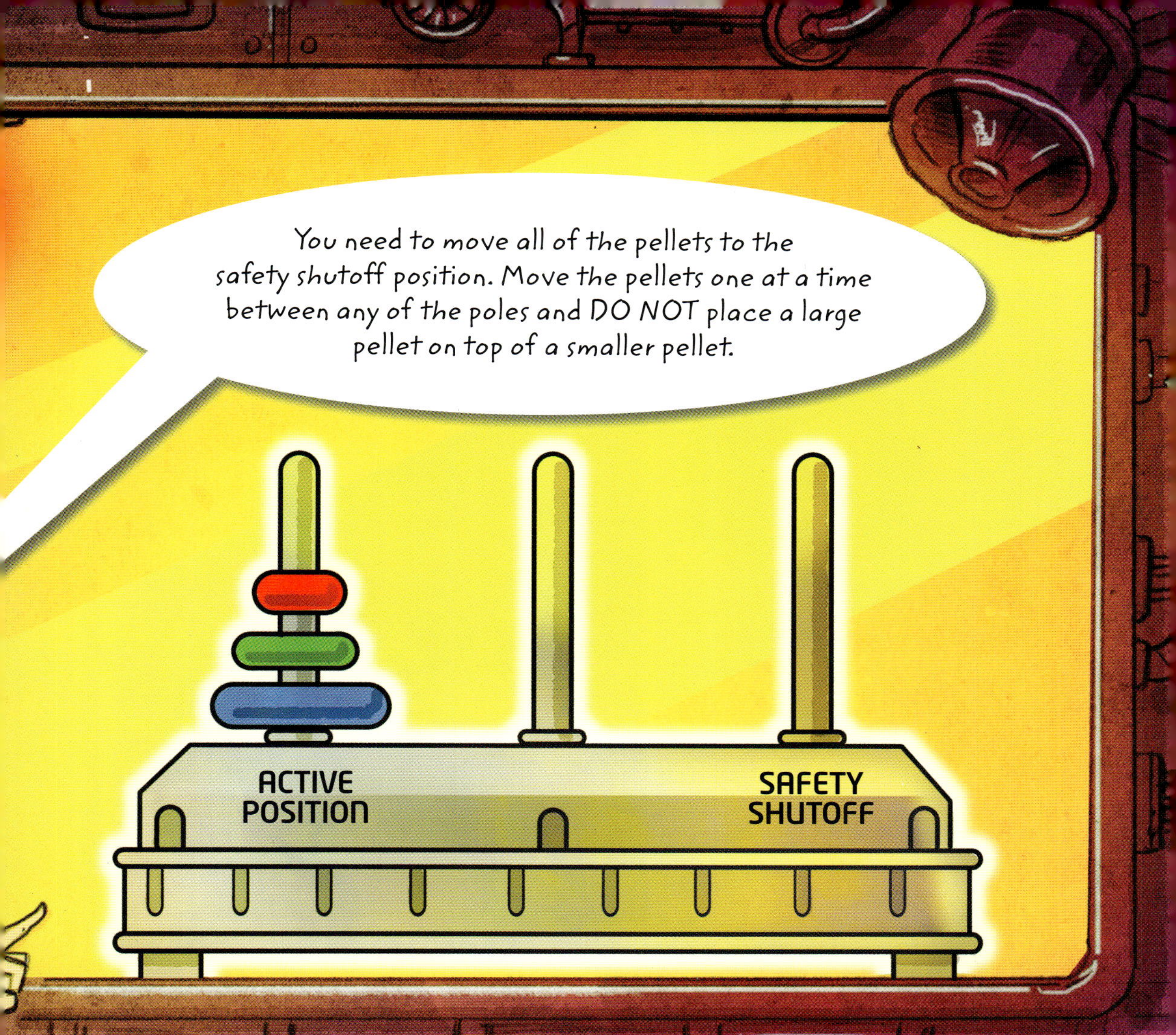

The reactor is too dangerous for you to crawl inside, but Superflip could do it.

What is the smallest number of moves that Superflip can make to activate the safety shutoff?

6	3	7
6 MOVES	**3 MOVES**	**7 MOVES**
HEAD OVER TO PAGE 16	FLIP TO PAGE 41	GO TO PAGE 30

No, you'll end up with two top squares and no bottom square.

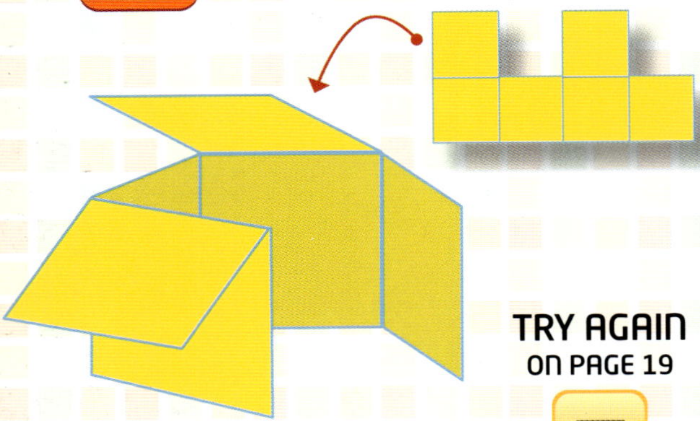

TRY AGAIN ON PAGE 19

That looks nothing like a square!

GO BACK TO PAGE 39 AND TAKE **ANOTHER LOOK**

You got it, but not quite all of it! Sunflower seed pods spiral to the left and right, using numbers from the Fibonacci sequence to squeeze as many seeds as possible into the flowerhead.

HAVE ANOTHER GO ON PAGE 13

Only an equilateral triangle (with equal sides) has rotational symmetry. It has a rotational symmetry of order 3, so the shape looks the same in 3 different positions.

GO BACK TO PAGE 10 AND **TRY AGAIN**

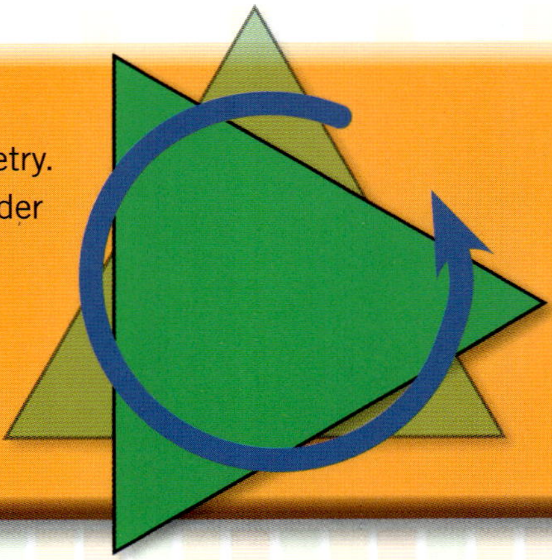

Stickman locks you in a dark room. You can see a triangle-shaped keyhole in the door. You look around to see what you could use as a key.

Which object will fit into the lock? Think about what happens if you cut each shape once.

A SQUARE GO TO PAGE 8

A CIRCLE TURN TO PAGE 39

A RECTANGLE FLIP TO PAGE 17

7

Wrong! Try multiplying together the first two numbers of each **row**. Can you see the pattern?

TURN BACK TO PAGE 19 AND TRY AGAIN 6

Wrong!
To get the next number, you need to add the two previous numbers together.

5 + 8 = ?

TRY AGAIN ON PAGE 37

Close, but 38 doesn't fit this pattern.

$$3 \times 8 + 3 + 8 = 35$$

TRY AGAIN ON PAGE 21

Wrong! $\frac{1}{2}$ would be 0.5 kg at each mealtime.

TRY AGAIN ON PAGE 33

These numbers are factors of 90, but 9 is not a prime number. Remember, prime numbers can only be divided by 1 and themselves.

GO BACK TO PAGE 12 AND TRY AGAIN

No, to make a cube you need six faces, not five.

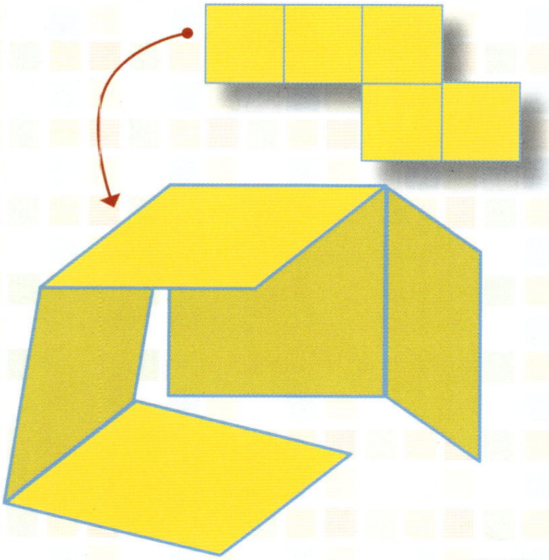

HAVE ANOTHER TRY ON PAGE 19

Wrong! Remember, not all of the triangles are the same size.

TURN BACK TO PAGE 43 AND LOOK A LITTLE CLOSER

Yes! A googolplex is a 1 followed by a 'googol' of zeros (1 googol = 1 with 100 zeros after it). There isn't enough room in the universe to write this number.

The security guard speaks into his radio, and ten minutes later a man appears. You tell him about the stolen Core and he instructs you to board his patrolship.

No! This special sequence of numbers is created by adding the two previous numbers to get the next number in the series.

HAVE ANOTHER GO ON PAGE 37

7

Disaster averted! You managed it in seven moves and you hardly broke a sweat. Being a hero is a piece of cake.

Well done, Superflip! He wags his tail happily. The Planet Star people want to throw a party in your honour, but you politely refuse and explain your mission.

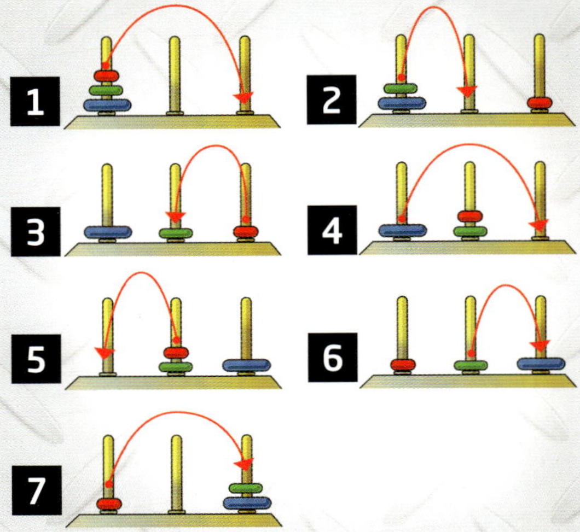

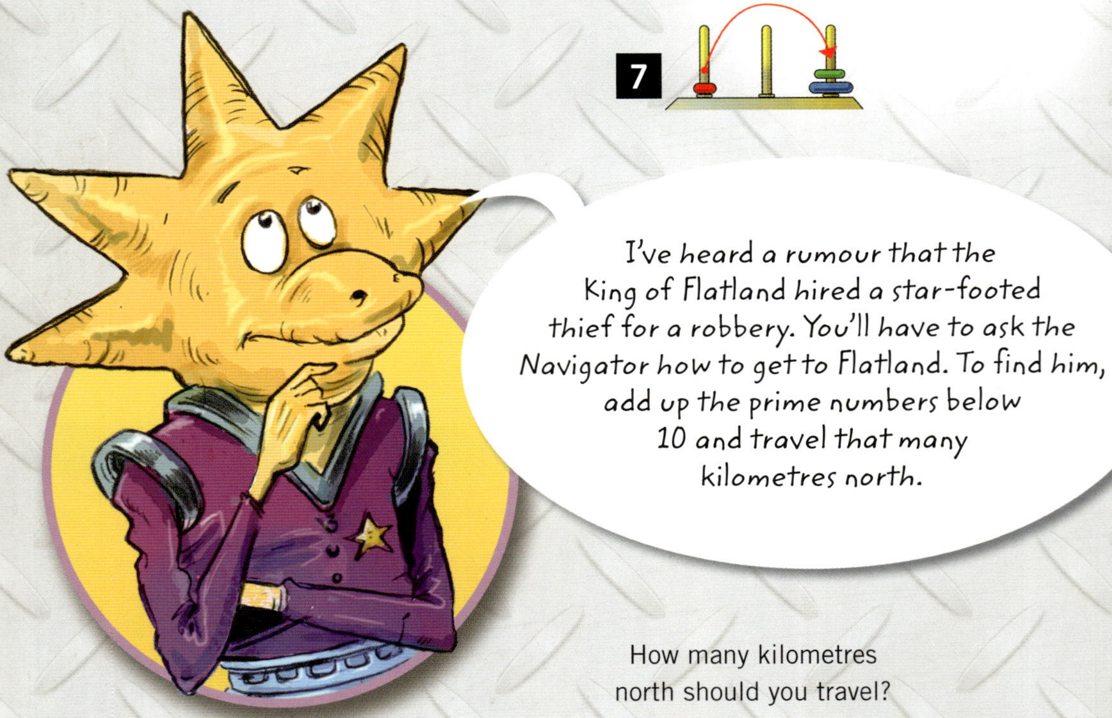

How many kilometres north should you travel?

2 + 3 + 5 + 7 = 17 KM
GO TO PAGE 9

2 + 4 + 6 + 8 = 20 KM
HEAD OVER TO PAGE 13

2 + 3 + 5 + 8 = 18 KM
FLIP TO PAGE 38

Correct! The snark-keeper jumps for joy and nearly crushes you with a tight hug.

$$\frac{3}{4} \times 9 \times 3 = 20\frac{1}{4} \text{ kg}$$

You're good with decimals AND fractions! What can I do to repay you? I know – I'll give you a quarter of the quarter of my chocolate cake...if you can tell me what a quarter of a quarter is!

Your brain is starting to hurt. But the cake does look good, and you're very hungry.

What is a quarter of a quarter?

$\frac{1}{2}$ TURN TO PAGE 12

$\frac{1}{16}$ TURN TO PAGE 16

You'd better check your sums.

$$88{,}122 \times 4 = 352{,}488$$

HAVE ANOTHER GO ON PAGE 29

A Rubik's Cube has 8 corner cubies, which are shared between 3 of the Cube's faces. Only cubies that do not run along the edges of the Rubik's Cube are not shared between its faces.

TURN BACK TO PAGE 8 AND TRY AGAIN

That's right – 43 quintillion possible positions! The door swings open.

Inside the room, the commanders are tied up next to the Rubik Computer. Professor Illogiko is holding the Core. You notice that he has a star-shaped peg leg – that explains the footprints!

Before you can answer, the Cube unfolds into a net (its 2D shape) and makes two copies of itself!

WHAT'S GOING ON?
FIND OUT ON PAGE 19

B

That's not right! Look closely at the numbers on the outside and the number in the circle. How do they work together?

$8 \times 9 \times 2 = 144$

TRY AGAIN ON PAGE 22

Fantastic! $\frac{1}{4} + \frac{9}{12}$ = one whole mooncheese! You and Superflip lead the snarks towards an enclosure with the wedges of cheese. When the last one is inside, you shut the gate and look for the snark-keeper.

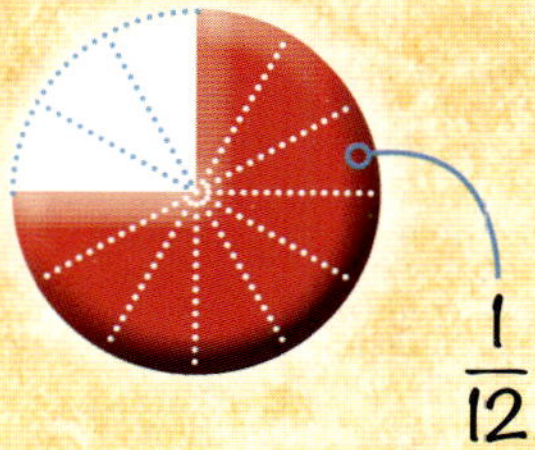

It doesn't take long to find him.

"You need to look after your snarks!" you tell him.

You love a maths challenge, so you take the feeding instructions from him.

What fraction of a kilogram does a snark eat at each mealtime?

$\frac{1}{3}$ **GO TO PAGE 8**

$\frac{3}{4}$ **TURN TO PAGE 20**

$\frac{1}{2}$ **JUMP TO PAGE 28**

20

Phew! There are 8 corner cubies (with 3 stickers each) and 12 edge cubies (with 2 stickers each) that move. As you click the final pieces into place, you rush towards the Core's power cables and reattach it before anyone can stop you.

The Rubik Computer fires up and you push the alarm button. Gates drop from the ceiling, blocking every door and window.

Within minutes, the police arrive, bundle an angry Illogiko into their van and free the commanders.

Congratulations, Lieutenant! You recovered the Core and no one can dispute your skills with a Rubik's Cube. You did disobey our order NOT to rescue us... But I guess we can overlook that and promote you to Captain! Congratulations!

No, that shape doesn't resemble your ship at all!

GO BACK TO PAGE 40
AND CHOOSE AGAIN

"You did it! Now tell me, why did you take my mosaic tile in the first place?"

You explain your mission to the King and tell him the rumours you heard about a star-footed thief.

"I work with Bernard Starfoot, a great antique finder, not a thief! But I know a way to help you."

The King unlocks a safe at the end of the room.

The All-seeing Cube, the only 3D object in Flatland, can answer any question. When you hold it, a symbol will appear on the top face. You may ask it where the Cube is. But first, you must say aloud which symbol is on the opposite face. This 2D pattern will help you. Try it now.

You hold the All-seeing Cube and an O shape appears on the top face. Which shape will be on the opposite face?

*** SHAPE** TURN TO PAGE 5

X SHAPE FLIP TO PAGE 22

= SHAPE HEAD OVER TO PAGE 10

 # SHAPE JUMP TO PAGE 41

Correct! **21,978 x 4 = 87,912.**
Weird, but true!

The patrolship starts moving so fast that your cheeks wobble. An alarm sounds – autopilot has malfunctioned. The ship won't stay on course!

The flight path numbers swirl across the navigation screen.

1... 1... 2... 3... 5... 8...

You need to take the controls and enter the next numbers. What do you type?

10, 13, 16
TURN TO PAGE 30

13, 21, 34
FLIP TO PAGE 13

9, 11, 14
GO TO PAGE 27

A square has rotational symmetry of order 4, so as it is rotated, there are 4 positions where it looks the same.

FLIP BACK TO PAGE 10 AND TRY AGAIN 9

Think again! If you were to cut $\frac{3}{4}$ into 9 pieces, the snarks will have less than $\frac{1}{9}$ each.

GO BACK TO PAGE 14 AND TRY AGAIN A

No, moving fences **C** and **E** leaves you with three squares and an L-shape.

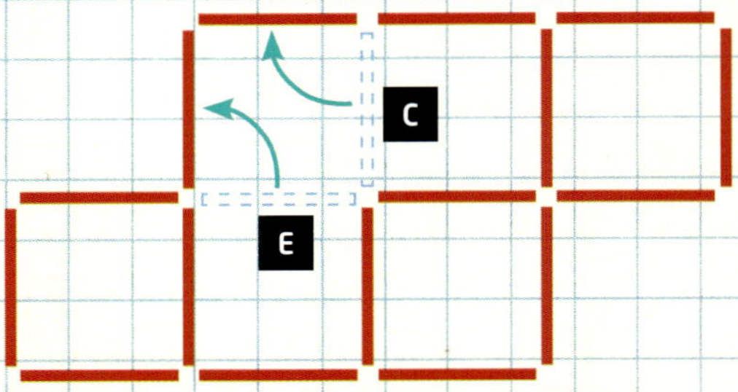

TRY AGAIN ON PAGE 9

No, 8 is not a prime number because it can be divided by 2 and 4. A prime number can only be divided by 1 or itself.

GIVE IT ANOTHER TRY ON PAGE 30

Wrong! The number pattern can be worked out by looking at the two numbers in triangles and the number in the circle.

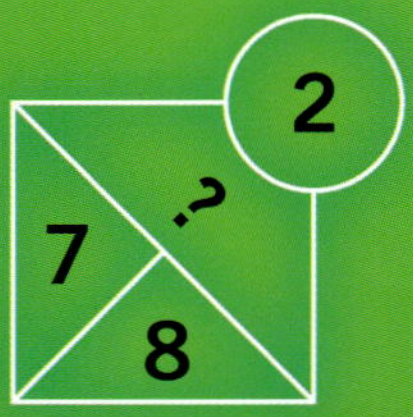

7 X 8 X 2 = ?

HAVE ANOTHER GO ON PAGE 22

A mollusc uses the Fibonacci sequence to build its spiral shell, allowing it to grow without changing its shape. But it's not the only living thing that uses it.

LOOK AT THE OPTIONS AGAIN ON PAGE 13

100 moves! You're supposed to be the best Cuber in the galaxy, so you'd better do it in fewer moves than that!

TRY AGAIN ON PAGE 41

No! With one cut, you can only make a semi-circle or a segment.

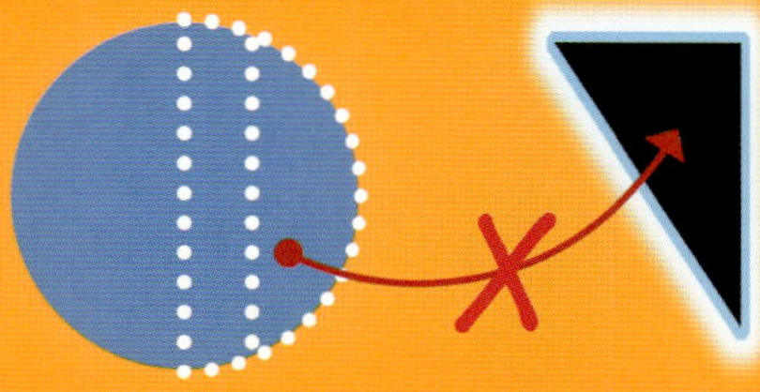

TRY AGAIN ON PAGE 27

When you reach the palace gates, Stickman pushes a button, but nothing happens.

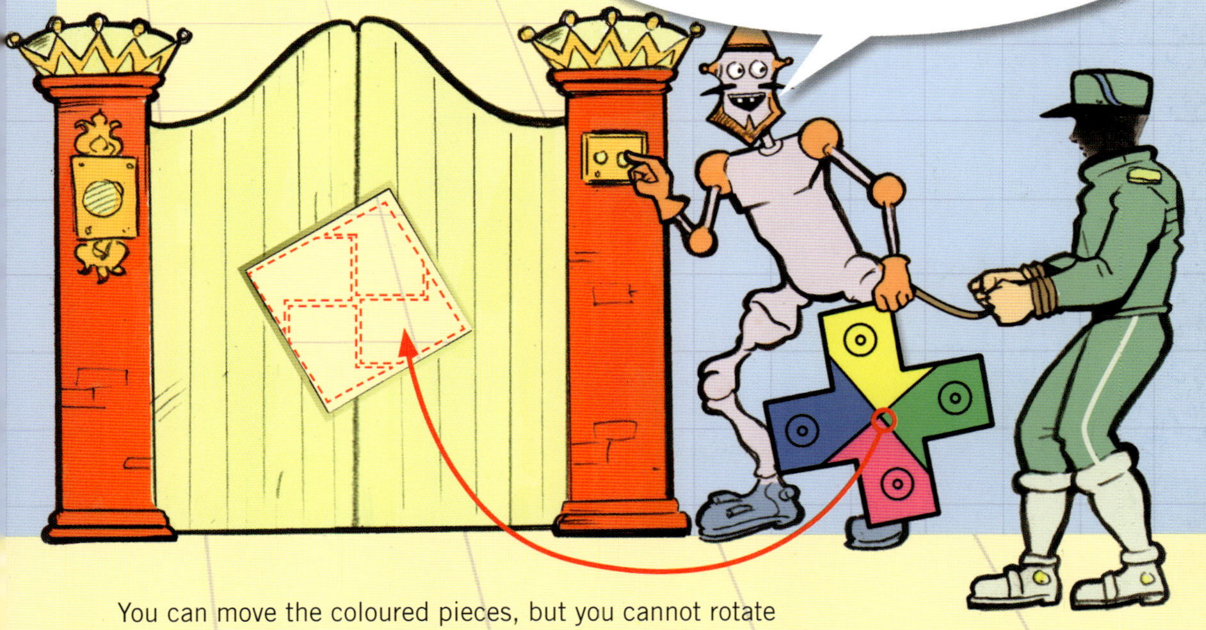

You can move the coloured pieces, but you cannot rotate them. How can the pieces make a square?

YELLOW TO THE RIGHT, GREEN TO THE BOTTOM, PINK TO THE TOP, AND BLUE TO THE LEFT.
GO TO PAGE 16

YELLOW TO THE BOTTOM, GREEN TO THE RIGHT, PINK TO THE LEFT, AND BLUE TO THE TOP.
FLIP TO PAGE 26

YELLOW TO THE BOTTOM, GREEN TO THE LEFT, PINK TO THE TOP, AND BLUE TO THE RIGHT.
TURN TO PAGE 43

Correct! These prime numbers will multiply together to give 90.

$3 \times 3 \times 2 \times 5 = 90$ ✔

You wave goodbye to the Navigator, board your ship and punch the numbers into your computer. You can see the Time Torus portal in the distance. As you get closer, it splits into 3 different-shaped portals.

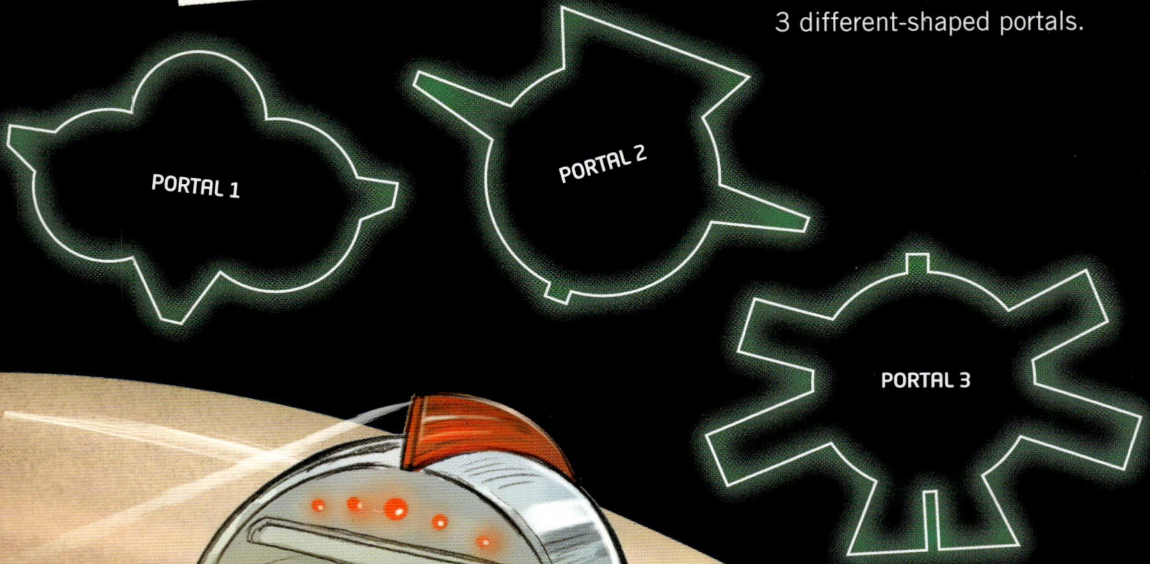

Which portal will your ship fit through?

PORTAL 1
GO TO PAGE 36

PORTAL 2
TURN TO PAGE 6

PORTAL 3
FLIP TO PAGE 23

Well done! The Cube reassembles itself.

Illogiko tells you to solve the Cube – NOW! You study the faces. You're just about to turn it, when…

THE CUBE CAN ALWAYS BE SOLVED IN 20 MOVES OR FEWER.
GO TO PAGE 21

THE CUBE CAN ALWAYS BE SOLVED IN 100 MOVES OR FEWER.
TURN TO PAGE 38

THE CUBE CAN ALWAYS BE SOLVED IN 33 MOVES OR FEWER.
FLIP TO PAGE 5

11 certainly doesn't fit this pattern.

1x1+1+1=3

TRY AGAIN ON PAGE 21

You're not moving the pellets correctly, because the puzzle cannot be completed in three moves.

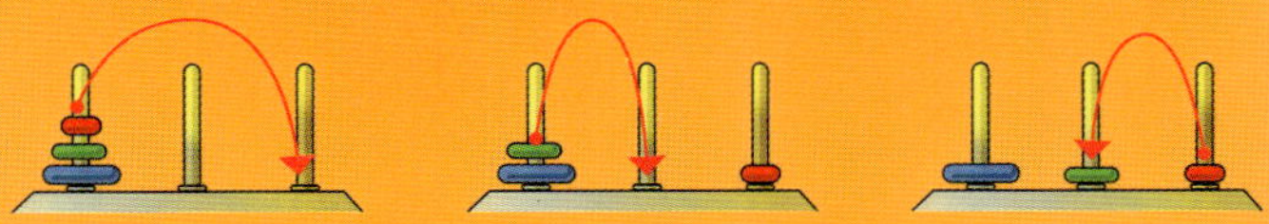

TURN TO PAGE 24 AND **START AGAIN**

Try again! The # symbol is not on the opposite side.

THINK AGAIN AND GO BACK TO PAGE 36

12

A cube has 12 edges, but only 6 faces. Cubies along edges and corners are shared between faces.

TURN BACK TO PAGE 8 AND TRY AGAIN

You're partly right! The cochlea in the inner ear is built using the Fibonacci sequence of numbers, which gives it a perfect spiral shape. However, these numbers are found in other places too.

TAKE ANOTHER LOOK ON PAGE 13

You reach Patrol HQ. A guard is standing at the security gates.

You may as well give it a go. What do you say?

A CENTILLION
GO TO PAGE 18

A GOOGOLPLEX
JUMP TO PAGE 29

A NONILLION
TURN TO PAGE 12

Well done. You've transformed the cross into a square and the hoverboard fits perfectly through the entrance.

You reach the front door, which is shaped like a triangle.

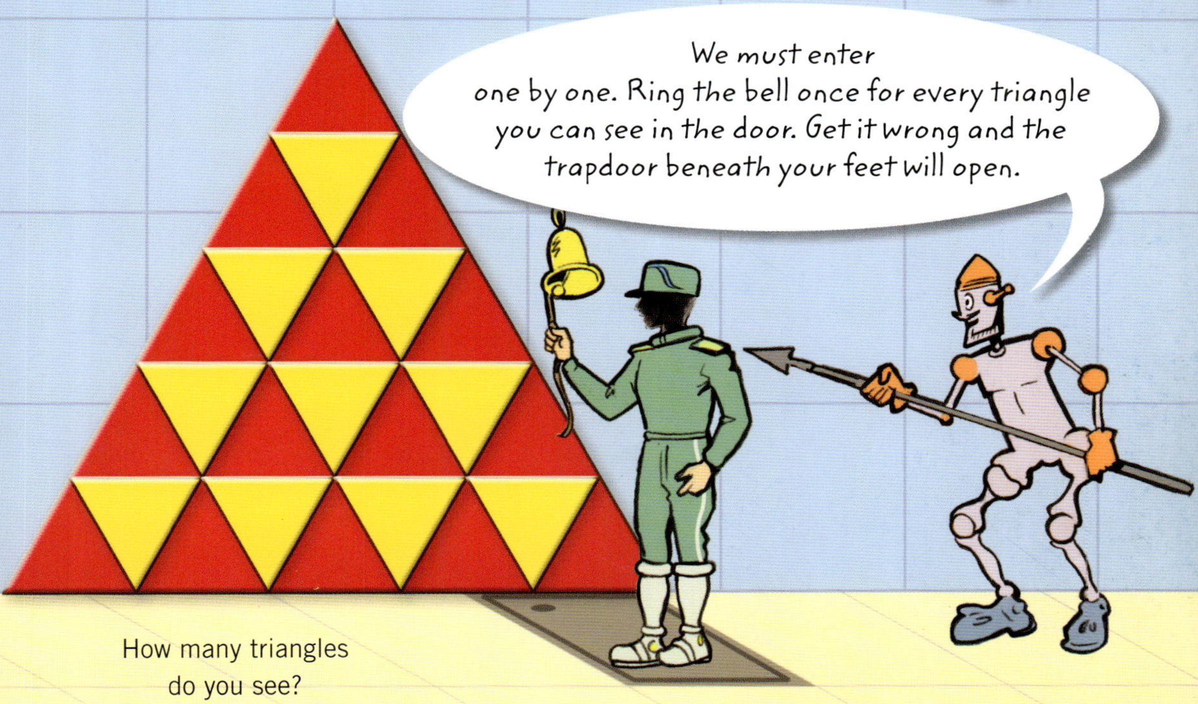

How many triangles do you see?

25 TRIANGLES
FLIP OVER TO PAGE 28

15 TRIANGLES
GO TO PAGE 20

48 TRIANGLES
TURN TO PAGE 11

Oh dear! Your maths ability has taken a turn for the worse.

Each snark eats $\frac{3}{4} \times 3 = 2\frac{1}{4}$ kg every day and there are only nine of them.

TRY AGAIN on PAGE 20

Uh oh! That piece is important. You'd better stop guessing.

TRY AGAIN on PAGE 17

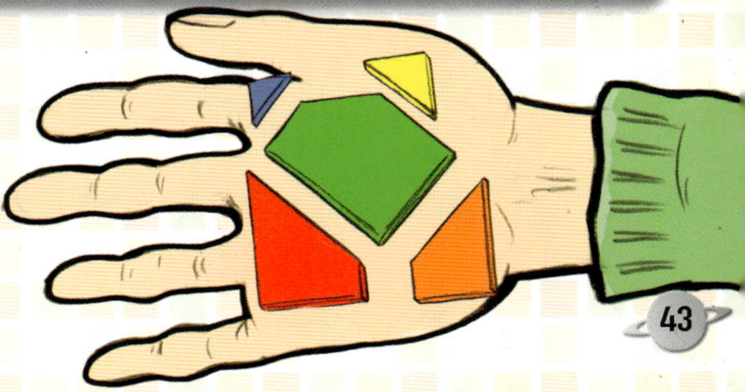

GLOSSARY

2D

Two-dimensional (2D) objects are flat shapes, just like the shapes on a sheet of paper. They can only be measured in two directions – length and width – and they have no depth or thickness.

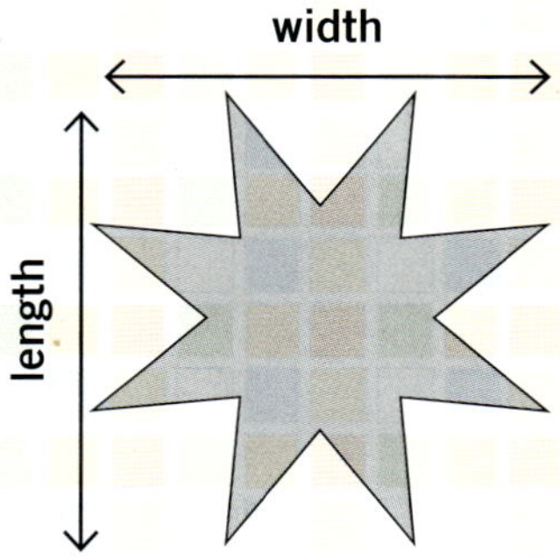

3D

Three-dimensional (3D) objects are solid things, like tables and chairs. They can be measured in three directions – length, width and depth.

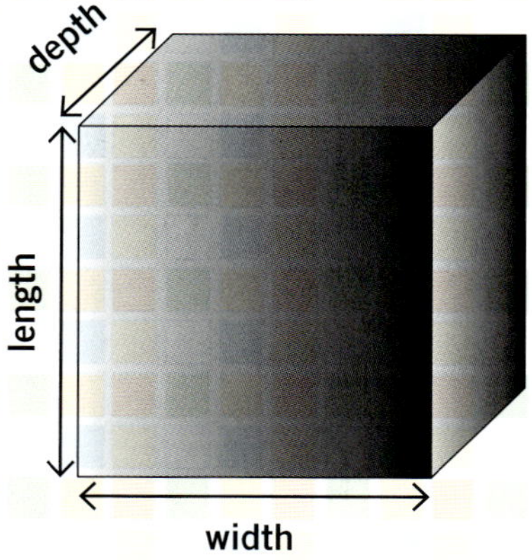

Coordinates

A set of numbers that tells you where something is located. Maps and GPS systems use a grid to pinpoint the coordinates of a place. Mathematicians use a similar system to plot the position of points on a graph.

Decimal

Decimal means 'made from groups of ten'. The decimal 15.5 has 1 ten, 5 units and 5 tenths. The digit after the decimal point is not a whole number, but a fraction. The decimal fraction 0.7 is the same as 'seven tenths' or $\frac{7}{10}$.

Everyone in Flatland is 2D!

Divide

To share something in equal parts. To 'divide something by five' means to split it into five equal parts.

Equilateral triangle

A triangle with sides of equal length.

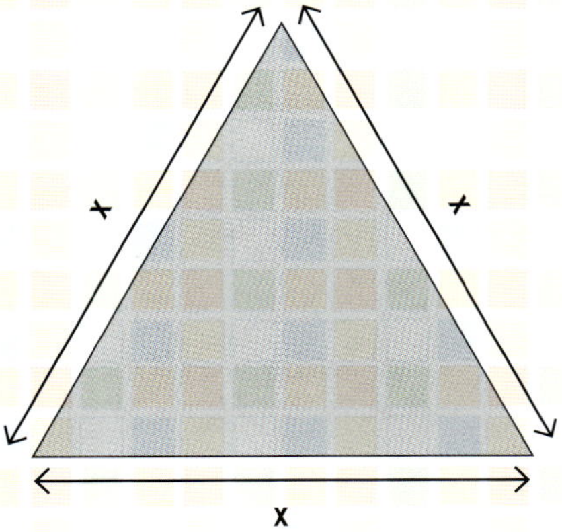

all sides of equal length

Factor

A factor of a number is any number that divides it exactly. Factors include 1 and the number itself. Prime factors are the prime numbers that divide a number perfectly.

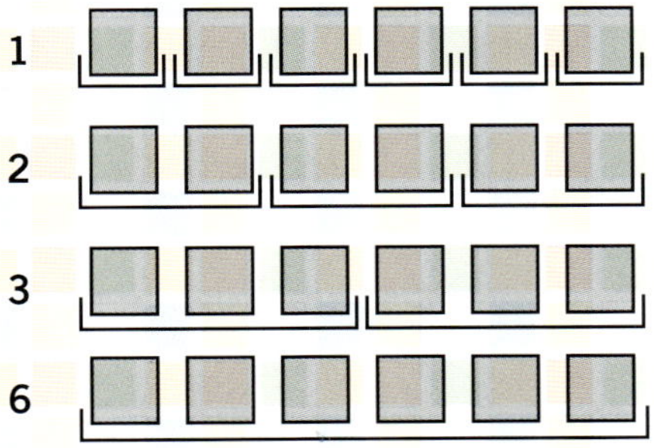

factors of 6

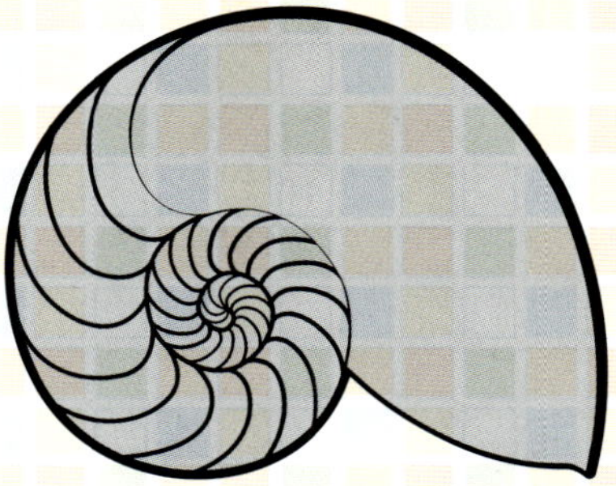

Fibonacci sequence

A special sequence of numbers in which the next number in the series is found by adding together the previous two. The Fibonacci sequence is regularly found in nature, as it is often the most efficient way of growing. It starts 1, 1, 2, 3, 5, 8, 13...

Fraction

A fraction is a number that is less than 1. Half is a fraction. Half a cake is less than a whole cake. The fraction $\frac{1}{4}$ means that the whole is divided into 4 equal parts and you have 1 of these parts. A fraction is written in this way:

$$\frac{1}{4}$$

1 — **numerator**

4 — ***denominator***

You can also write fractions as decimals. To convert a fraction into a decimal fraction, divide the numerator by the denominator, like this:

$\frac{1}{4} = 1 \div 4 = 0.25.$

God's number

The minimum number of moves that you need to solve any Rubik's Cube. Every position of the Rubik's Cube can be solved in 20 moves or fewer.

Isosceles triangle

A triangle that has two sides of equal length.

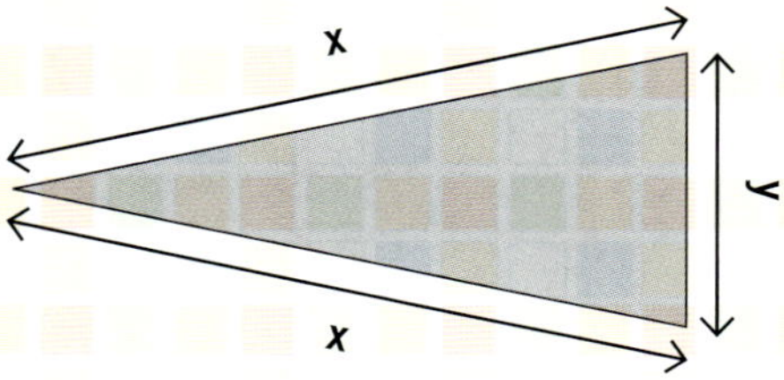

two sides of equal length

Kilogram

A metric measure of weight. A bag of flour weighs 1 kilogram (kg).

Multiply

A way of grouping together numbers. Multiplying a number adds it to itself a certain number of times. For example: 3 x 4 is the same as 3 + 3 + 3 + 3.

Navigation

Another word for finding a route. Navigation means finding a location by plotting your way around the neighbourhood, world or universe.

Net (cube)

The flat 2D shape that can be folded to make a 3D object, such as a cube.

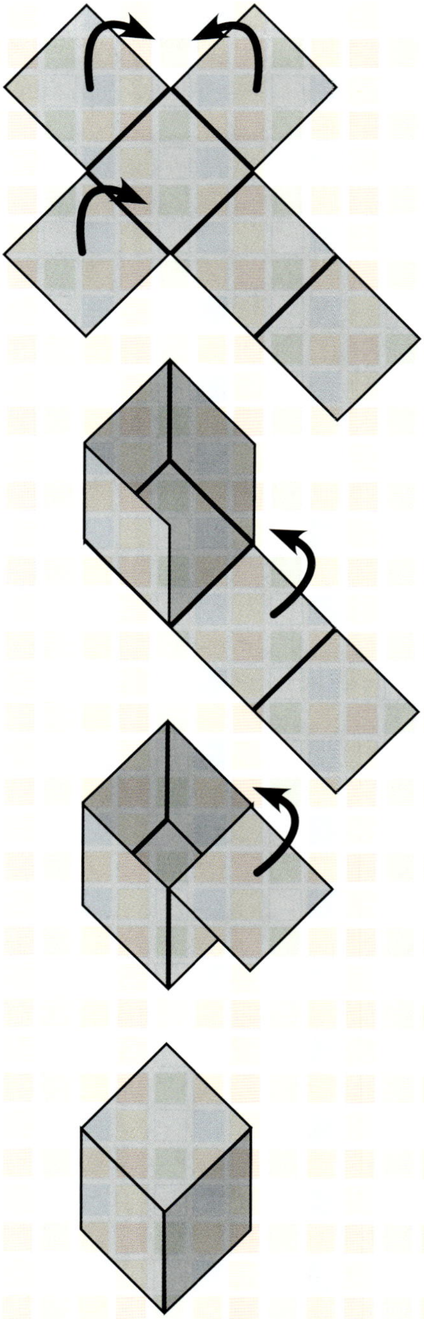

Pattern
A design that is repeated; a regular arrangement of shapes or numbers.

Portal
A doorway to somewhere else.

Prime number
A number that can only be divided by itself and 1 (these are its only factors). The first prime numbers are 2, 3, 5, 7 and 11.

Reactor
A container for chemical or nuclear reactions to release energy. Reactors are often reinforced so that they don't explode.

Rotational symmetry
Some shapes look exactly the same when you turn them. If a shape fits into itself more than once as you rotate it, then it has rotational symmetry.

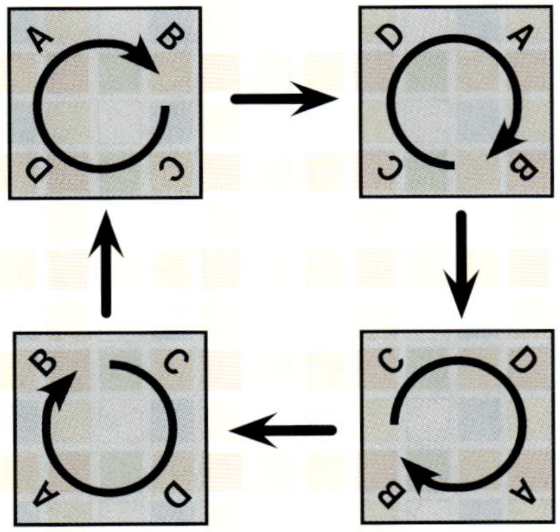

a square has a rotational symmetry of four

Scalene
A triangle that has no sides of equal length. All three sides are different lengths.

Sequence
A series of numbers that has a pattern. All sequences have a hidden rule which makes them work.

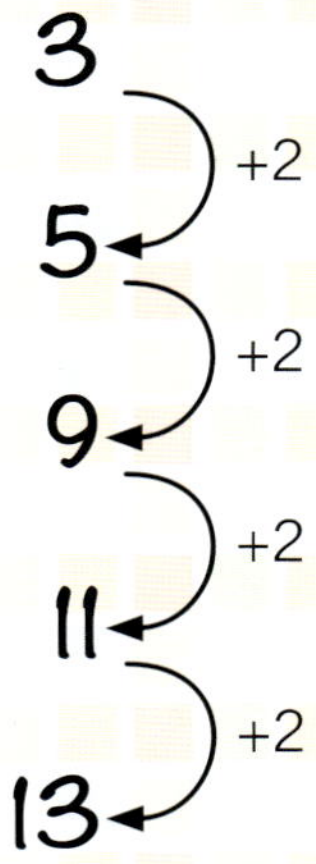

Symbol
A letter or image that stands for something else.

Taking it further

The Rubik's Quest series is designed to motivate children to develop their Science, Technology, Engineering and Mathematics (STEM) skills. They will learn how to apply their know-how to the world through engaging adventure stories involving the Rubik's Cube, a mind-bending puzzle used throughout the world by people of all ages. For each book, readers must solve a series of problems to make progress towards the exciting conclusion.

The books do not follow a conventional pattern. The reader is directed to jump forwards and backwards through the book according to the answers they give to the problems. If their answers are correct, the reader progresses to the next part of the story; if they are incorrect, the reason is explained before the reader is directed back to try the problem again. Additional support may be found in the glossary at the back of the book.

To support your child's development you can:

- Read the book with your child.
- Solve the initial problems and discover how the book works.
- Continue reading with your child until he or she is using the book confidently, following the **"GO TO"** instructions to find the next puzzle or explanation.
- Encourage your child to read on alone. Ask: "What's happening now?" Prompt your child to tell you how the story develops and what problems they have solved.
- Discuss shapes, patterns and numbers in everyday contexts. For example, look at fractions and decimals when dividing up a cake; look for patterns in nature, such as symmetrical shapes in flowers and leaves; learn about weights and measures in recipes.
- Have fun making up number sequences and patterns. Count in 2s, 3s, 4s and larger steps. List doubles, halves, primes and factors.
- Play card games involving numbers, such as snap or pairs.
- Play computer games that involve maths or shapes. These will hold children's interest with colourful graphics and lively animations as they practise basic number skills.
- Most of all, make learning fun!